IT'S A CHILD'S WORLD

Bugs and Other Insects

Using Nonfiction to Promote Literacy Across the Curriculum

by Doris Roettger

Fearon Teacher Aids
Simon & Schuster Supplementary Education Group

Teacher Reviewers

Rebecca Busch
San Antonio, Texas

Nora Forester
San Antonio, Texas

Lupita Guerrero
Houston, Texas

Delphine Hetes
Detroit, Michigan

Debbie Kellogg
West Des Moines, Iowa

Perry Kellogg
West Des Moines, Iowa

Ilene Schwartz
Houston, Texas

Editorial Director: Virginia L. Murphy

Editor: Virginia Massey Bell

Copyeditor: Susan J. Kling

Illustration: Anita Nelson

Design: Terry McGrath

Cover Design: Lucyna Green

Entire contents copyright © 1991 by Fearon Teacher Aids, part of the Education Group of Simon & Schuster, a Paramount Communications company. However, the individual purchaser may reproduce designated materials in this book for classroom and individual use, but the purchase of this book does not entitle reproduction of any part for an entire school, district, or system. Such use is strictly prohibited.

ISBN 0-86653-992-1

Printed in the United States of America

1.9 8 7 6 5 4 3 2

A Note from the Author

Children have a natural curiosity about the world in which they live. They are intensely interested in learning about real things, real places, and real people. They also enjoy and learn from hands-on experiences. Nonfiction books and magazines provide opportunities for children to explore their many interests and extend their base of knowledge.

Reading nonfiction materials is different from reading picture or storybooks. To be effective readers, children need to learn how to locate information or find answers to their many questions. They also need to learn to think about and evaluate the accuracy of any information presented. Finally, they need opportunities to learn the relationship between what they read and the activities in which they apply their new knowledge.

You, as the teacher, can provide opportunities for children to learn from their observations, their reading, and their writing in an integrated language-arts approach across the curriculum.

Modeling thinking strategies and then providing practice across the curriculum will help students become observers and explorers of their world, plus effective users of literacy skills. Encouraging children to extend and demonstrate their understanding through a variety of communication areas—speaking, reading, drama, writing, listening, and art—is also very valuable.

The suggestions in this guide are action-oriented and designed to involve students in the thinking process. The activities do not relate to any one single book. Instead, the strategies and activities are designed to be used with any of the books suggested in the bibliography or with books found in your own media center. The suggested interdisciplinary activities can also be used across grade levels.

Each lesson begins with the reading of a nonfiction book, book chapter, or magazine article—any title that relates to the follow-up activities. During the activity phase and at other class times, students are encouraged to return to the nonfiction selections available in the classroom to find answers to their questions, compare and verify their observations, and add any new information to their current knowledge base.

The individual theme units are designed to be used for any length of time—from a few days to a month or more, depending on the needs and interests of your students.

Suggested goals for this unit are provided near the beginning of this guide on page 16. The webs on pages 7-9 give you an overview of the areas in which activities are provided.

On each page of this guide, there is space for you to write reflective notes as well as ideas that you want to remember for future teaching. This guide is designed to be a resource from which you make decisions and then select the learning experiences that will be most appropriate for your students.

Doris Roettger

Contents

Literacy Skills 7

Interdisciplinary Skills 8

Learning and Working Strategies 9

About Bugs and Other Insects 10

Suggested Reading Selections 12

Instructional Goals 16

Getting Started
    What Do We Already Know? 18
    Cooperative Working Groups 22
    Finding Information in Books and Magazines 24

Real-Life Laboratory
    Collecting Insects 28
    Classroom Observation Centers 32
    Observe and Record 36

Cross-Curriculum Activities
    Writing Arena 40
    Art All Around 42
    The Drama Scene 46
    Fingerplays and Poetry 50

Cooking Makes It Memorable 52

Social Studies from Here to There 54

Science Sector 56

Membership Cards for a Bug Club Reproducible 58

Writing Helpers Reproducibles 59-61

Ant Pattern Reproducible 62

Cricket Pattern Reproducible 63

Grasshopper Pattern Reproducible 64

Literacy Skills

*T*he following literacy skills are addressed in the *Bugs and Other Insects* theme guide.

Vocabulary
1. Use new vocabulary in speaking activities
2. Use new vocabulary in writing activities

Writing
1. Record observations
2. Maintain a journal
3. Write paragraphs for a newsletter
4. Classification

BUGS AND OTHER INSECTS

Reading
1. Read a variety of books and magazine articles
2. Read to find answers to questions
3. Read poetry

Listening, Speaking
1. Listen for answers and new information
2. Listen to student presenters
3. Chant poetry
4. Fingerplays
5. Describe an insect's life

Interdisciplinary Skills

*T*he following interdisciplinary skills are addressed in the *Bugs and Other Insects* theme guide.

Science

1. Collect insects
2. Create observation centers
3. Observe insects
4. Record observations
5. Develop a food chain

Art

1. Create a habitat mural
2. Put insects together
3. Build a butterfly mobile
4. Construct magnetic ants

BUGS AND OTHER INSECTS

Cooking

1. Create butterflies from stuffed celery and pretzels
2. Ants on a log
3. Bake bug-shaped cookies

Drama

1. Act out an insect in a narrative pantomime
2. Describe an insect's life

Social Studies

1. Use a map to trace migration routes
2. Use a globe to trace migration routes

Learning and Working Strategies

The following learning and working strategies are addressed in the *Bugs and Other Insects* theme guide.

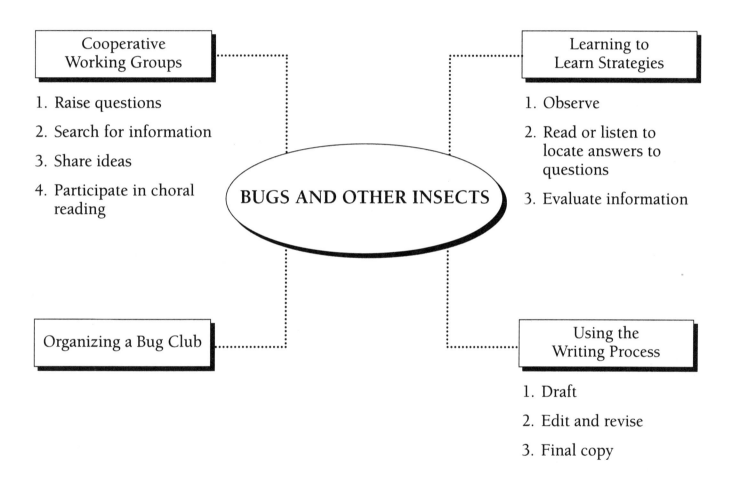

Cooperative Working Groups

1. Raise questions
2. Search for information
3. Share ideas
4. Participate in choral reading

Organizing a Bug Club

Learning to Learn Strategies

1. Observe
2. Read or listen to locate answers to questions
3. Evaluate information

Using the Writing Process

1. Draft
2. Edit and revise
3. Final copy

About Bugs and Other Insects

We often use the words *bugs* and *insects* interchangeably, but actually bugs are specific insects with particular attributes, such as beaks for piercing and sucking. Also, some bugs secrete substances, while others (like the stinkbug) may emit unpleasant odors. For more information about bugs, you and your students will want to look up the insect order *Hemiptera* in an encyclopedia or other reference material.

Insects live almost everywhere on earth—from tropical jungles to very cold regions, from mountains to deserts. Insects are so numerous that in one square mile they can equal the total number of people on the planet earth.

Most insects have three body parts and six legs, although they can be many different sizes, shapes, and colors. Many adult insects have wings. Most insects are less than one fourth of an inch long. However, some, like the walking stick, may measure up to four inches. Some insects blend in with their surroundings, looking like sticks, green leaves, dried leaves, or thorns. Others may be very flamboyant with bright, jewel-like colors.

Some insects have large compound eyes that occupy most of their heads. A compound eye is made up of tiny six-sided lenses. Insects, however, cannot move or focus their eyes. Some insects do not have any eyes at all.

Most insects do not have ears. They hear through delicate hairs on their antennae or on other parts of their bodies. Insects are very sensitive to touch. They can feel through the hairs and spines that cover their body parts. Insects breathe through tiny holes along the sides of their bodies.

Many insects have a much stronger sense of taste than we do. They can taste sweet, sour, salty, and bitter flavors. Some insects, like ants, bees, and wasps, taste through their antennae. Others, such as butterflies, some moths, flies, and honeybees, taste with their feet. Insects do not have a voice, but they may make sounds by rubbing one body part against another.

Insects can be very useful to all of us. Bees, wasps, flies, butterflies, and moths pollinate plants. Many insects are also food for frogs, lizards, birds, fish, skunks, and other animals. A few insects are harmful to people. Some may bite us, while others eat crops or cause damage in other ways. Many insects do the same things that people do. They build houses, bridges, and even raise crops. Some have jobs in their own insect communities, such as nurses, soldiers, and hunters.

Caterpillars don't fit the description of an insect—that is, having three body parts and six legs. However, a caterpillar is the larva stage of a butterfly or moth, both of which are insects. A caterpillar has thirteen segments, three pairs of true legs, and as many as five pairs of soft, false legs. Caterpillars are plant eaters with enormous appetites, which is why they are often thought of as pests.

Common insects your students may easily find and the foods they eat are listed here for your convenience.

Insects	**Some Foods They Eat**
ants	crumbs
bees	honey, nectar
monarch butterflies	sweet nectar
crickets	lettuce, apples, crumbs
caterpillars	leaves (but different types of caterpillars eat different leaves)
flies	any type of food
fireflies	don't eat (sprinkle water on them)
grasshoppers	fresh grass
ladybugs	aphids
moths	nectar, sap from trees, water
wasps	nectar, pollen, honey

Suggested Reading Selections

A variety of nonfiction and fiction selections for the primary grades is suggested for use with this theme unit. You will probably want to assemble a collection of materials ahead of time. Or, you may wish to have the students help collect several titles from the library as a group activity. The number and type of selections you and the children read will depend on the length of time you devote to this unit, as well as the availability of titles and the level of your students.

Nonfiction Picture Books

Ants Are Fun by Mildred Myrick. New York: Harper & Row Publishers, 1968. A Science I Can Read Book. A good description of ants and how they live, plus directions on how to build an ant nest or city.

Ant Cities by Arthur Dorros. New York: Harper & Row Publishers, 1987. A Let's Read-and-Find-Out Book. Explains how to build an ant city. Includes color illustrations of ants and their colonies and a good description of an ant's life.

Backyard Insects by Millicent Selsam and Ronald Goor. New York: Four Winds Press, 1981. Discusses common garden insects and their protective appearances, including camouflage, warning colors, scary appearances, and copycat characteristics.

The Bee by Iliane Roels. New York: Grosset & Dunlap, 1969. Color pictures, together with simple text, provide a good description of the responsibilities of different bees and how new hives are formed.

Bugs by George McGavin. New York: The Bookwright Press, 1989. From the Discovering Nature series. Large print with lots of details, color illustrations, key vocabulary in bold, table of contents, glossary, and index.

Bugs by Nancy Winslow Parker and Joan Richards Wright. New York: Greenwillow Books, 1987. Excellent labeled, color illustrations with double page for each insect. Left side provides question with answer, right side gives information about the insect.

Bugs, Bugs, Bugs by Sandra Granseth and Diana McMillen. Des Moines, Iowa: Meredith Corporation, 1989. An arts and crafts cooking idea book.

Butterflies and Moths by Keith Porter. New York: The Bookwright Press, 1986. From the Discovering Nature series. Large print with lots of details, color illustrations, key vocabulary in bold, table of contents, glossary, and index.

Catch a Cricket by Carla Stevens. Reading, Massachusetts: Addison-Wesley Publishing Co., 1961. Provides information for catching and caring for crickets, grasshoppers, fireflies, and other creatures.

Discovering Bugs by George McGavin. New York: The Bookwright Press, 1989. From the Discovering Nature series. Large print, highlighted key vocabulary, color photographs, table of contents, glossary, and index.

From Egg to Butterfly by Marlene Reidel. Minneapolis: Carolrhoda Books, Inc., 1981. Goes through the four stages with good color illustrations.

Insect Pets: Catching and Caring for Them by Carla Stevens. New York: Greenwillow Books, 1978. Suggests ways to capture a variety of insects. Lists the materials needed, how to make a home for the insects, and ideas to find out further information.

Insects by Jeanne Brouillette. Chicago: Follett Publishing Co., 1963. Clearly labeled color illustrations, vocabulary list.

It's Easy to Have a Caterpillar Visit You by Caroline O'Hagan. New York: Lothrop, Lee & Shepard Books, 1980. Color illustrations show stages from caterpillar to butterfly.

Lucky Ladybugs by Gladys Conklin. New York: Holiday House, 1968. Simple text gives the entire life of a ladybug in a narrative format.

Let's Find Out About Insects by David C. Knight. New York: Franklin Watts, Inc., 1967. Simple text provides good, concise descriptions of a large number of insects, two-color illustrations, body parts labeled.

Life of the Butterfly by Heiderose and Andreas Fischer-Nagel. Minneapolis: Carolrhoda Books, Inc., 1987. Beautiful color, detailed pictures, four stages well-described, glossary, and index.

Monarch Butterflies by Emilie Lepthien. Chicago: Childrens Press, 1989. Beautiful color pictures and a good description of each stage of the monarch butterfly's life.

Poetry and Fingerplays

Hand Rhymes by Marc Brown. New York: E. P. Dutton, 1985. Fourteen hand rhymes, including "Where Is the Beehive?" and "The Caterpillar."

Joyful Noise by Paul Fleischman. New York: Harper & Row Publishers, 1988. Two-voice poetry, good for choral reading.

Fiction Picture Books

Bug City by Dahlov Ipcar. New York: Holiday House, 1975. In Bug City, the bug family members sleep in bedbugs, use a sawfly, look at the latest bookworms, wear yellow jackets, and buy potato bugs.

The Grouchy Ladybug by Eric Carle. New York: T. Y. Crowell Co., 1977. A delightful book in which a grouchy ladybug goes looking for a fight and challenges everyone she meets.

Two Bad Ants by Chris Van Allsburg. Boston: Houghton Mifflin, 1988. When two bad ants desert their colony, they experience a dangerous adventure that convinces them to return to their former safety.

The Very Hungry Caterpillar by Eric Carle. New York: Philomel Books, 1981. A caterpillar eats his way through different fruits before turning into a butterfly in this memorable counting book.

Teacher Reference

Sharing the Joy of Nature: Nature Activities for All Ages by Joseph Cornell. Nevada City, CA: Dawn Publishing, 1988. A parents' and teachers' nature-awareness guidebook.

Magazine Articles

"Ants" from *Ranger Rick*, National Wildlife Federation, May 1988.

"Ants" from *Ranger Rick*, National Wildlife Federation, June 1989.

"Bees" from *Ranger Rick*, National Wildlife Federation, November 1988.

"Bees" from *Ranger Rick*, National Wildlife Federation, November 1989.

"Butterfly" from *Ranger Rick*, National Wildlife Federation, August 1988.

"Butterfly" from *Ranger Rick*, National Wildlife Federation, May 1989.

"Butterfly" from *Ranger Rick*, National Wildlife Federation, October 1989.

"Butterfly" from *Ranger Rick*, National Wildlife Federation, December 1989.

"Butterfly World" from *Ranger Rick*, National Wildlife Federation, December 1989.

"Caterpillar" from *Ranger Rick*, National Wildlife Federation, August 1988.

"Caterpillar" from *Ranger Rick*, National Wildlife Federation, December 1989.

"Firefly" from *Ranger Rick*, National Wildlife Federation, June 1988.

"Grasshopper" from *Ranger Rick*, National Wildlife Federation, August 1989.

"The Many Jobs of Buzz the Bee" from *Ranger Rick*, National Wildlife Federation, May 1990.

"Mosquito" from *Ranger Rick*, National Wildlife Federation, August 1988.

"Mosquito" from *Ranger Rick*, National Wildlife Federation, May 1989.

"Night Flyer" from *Ranger Rick*, National Wildlife Federation, March 1989.

Instructional Goals

*I*nstructional goals for this theme unit are provided here. Space is also provided so that you may fill in your own individual goals where appropriate as well. By the end of this theme unit, students should be able to:

1. Name all they know and think they know about insects.
2. Observe insects as they move and eat.
3. Distinguish between fiction and nonfiction selections.
4. Identify ants, bees, wasps, crickets, grasshoppers, caterpillars, and butterflies.
5. Write about their observations.
6. Raise questions about insects based on their own curiosity.
7. Learn to locate information using a card catalog or microfiche, table of contents, and index.
8. Find information to answer their questions.
9. Evaluate information they find and decide whether it is the information they wanted to know.
10. Share what they have learned with others.
11. Use new vocabulary in their writing and speaking.
12. Share what they have learned through drama, art, and writing.
13. Extend their learning through cooking.
14. ....
15. ....
16. ....
17. ....
18. ....
19. ....
20. ....

Getting Started

What Do We Already Know?

The following activities are designed to help launch the *Bugs and Other Insects* theme unit. You may want to use all of the activities or only one or two, depending on the needs of your students. At the beginning of each lesson, reading a nonfiction book or magazine selection to the class serves as a motivator and helps students become more familiar with and involved in using nonfiction selections. For the activities in this section, a general selection on insects would probably be most appropriate. You'll also want to provide plenty of opportunities for children to return to nonfiction selections independently during the activity phases and at other times during class periods as well.

1. Finding Out What Children Already Know

NOTES

Find out what the children already know about insects and what their misperceptions and questions are so that you can help direct their learning.

a. Ask children to name what they think they know for sure about insects. Write their suggestions on a large sheet of paper.

b. List what the children think they know about insects, but are not certain about, in a second column.

c. Ask children to raise questions they have about insects and list the questions in a third column. Leave enough space so that you can record possible sources to check for information.

d. As a class, decide which questions the group would like to answer.

Facts We Know About Bugs and Other Insects	What We Think We Know	Questions We Have and Things We Wonder About	What We Have Learned

2. Vocabulary

a. List the following vocabulary words on chart paper and then display the chart in the classroom. Display pictures of a variety of insects, too. As you use the vocabulary words during discussions about insects with the class, point out examples of the vocabulary words on the chart and in the displayed pictures as well.

b. Have children make their own flashcards of the vocabulary words. The flashcards can be used in practicing identification of insect body parts and in the students' writing.

camouflage	life cycle
molting	antennae
larvae	head
cocoon	thorax
pupa (chrysalis)	abdomen
metamorphosis	wing

NOTES

3. Organizing a Bug Club

Organize a bug club. As a group, decide what children need to do to join the club and when, where, and how often the club will meet. Discuss whether the club should elect leaders as well. Suggested activities for bug club meetings include:

a. making membership cards
   (see page 58 for patterns)

b. sharing information children have found

c. collecting insects

d. observing insects

e. art activities

f. poetry readings

g. fingerplays

h. making bug-shaped cookies

i. drama activities

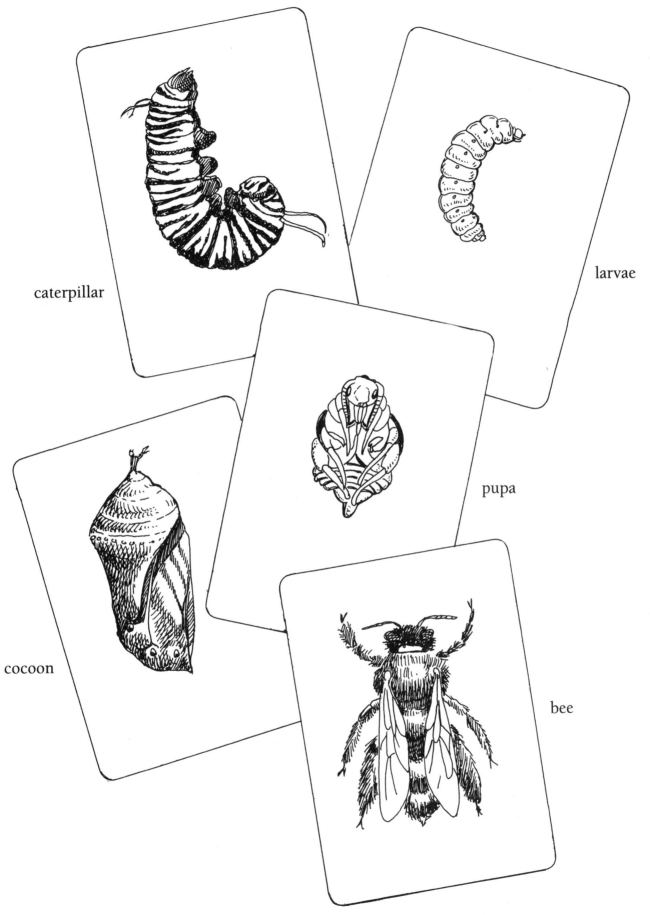

caterpillar

larvae

pupa

cocoon

bee

BUGS AND OTHER INSECTS **21**

Cooperative Working Groups

*H*ave children work together in small groups to share information, help one another, and learn how to function cooperatively as a team. Before beginning one of the activities in this section, share one or more nonfiction selections with the class. You might read an interesting paragraph about each of the insects listed in the first activity to get children thinking about which insects they wish to study in greater detail.

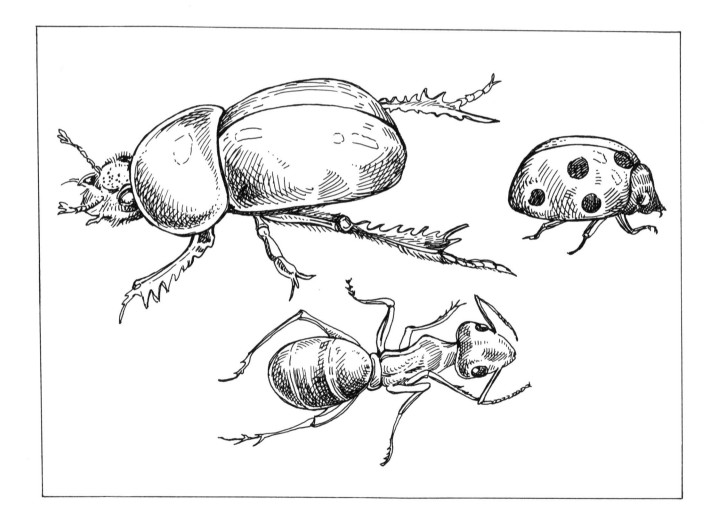

1. Forming Working Groups

Form working groups according to students' interests in the following insects:

a. ants
b. caterpillars and butterflies
c. crickets
d. grasshoppers
e. bees and wasps
f. ladybugs
g. other insects the class has found

NOTES

2. Raising Questions

Help groups raise "I wonder" questions. These questions can be written on large sheets of paper and displayed in the classroom. For example,

"I wonder what ants eat?"

"I wonder how caterpillars turn into butterflies?"

"I wonder how far grasshoppers can jump?"

"I wonder how crickets make their chirping sound?"

3. Locating Reading Selections

Help groups find a large variety of nonfiction and fiction reading selections to set up in the classroom library or in an interest center. Suggested selections are found on pages 12-15 in this teacher guide.

4. Working Together

Groups can work together to:

a. search for answers to their questions.

b. share their ideas with other students.

c. participate in choral reading activities.

d. present their information at bug club meetings.

e. participate in drama activities.

f. write a book of facts based on what they have learned.

g. share what they have learned through art.

Finding Information in Books and Magazines

*F*or each of the following activities, first select a nonfiction book or magazine and then demonstrate how to find information by thinking aloud and having children work through the process with you. Share what you are doing so that children can learn the thought processes as they learn the strategies. Gradually, ask students to think aloud as they locate answers to their own questions as they read as well.

1. Locating Information

a. Help children find books and magazines that pertain to an insect they are researching.

b. For more advanced readers, model ways children can use a table of contents. Think aloud so students know what you are doing. Then give groups practice identifying titles, page numbers, and locating the pages in books and magazines. Have children think aloud so you know that they understand the process. Repeat the process using an index.

2. Finding Answers to Questions

a. Demonstrate how students can find answers to questions by modeling how you would locate an answer to a question posed by one of the children. Again, think aloud as you find the answer. Give students practice in finding an answer to a question. Ask them to think aloud as they find their answers.

b. Encourage students to read books of their choice or have older students read several books to a cooperative working group. You might also make tapes of appropriate selections to place in a listening center. Encourage children to watch or listen for new and interesting facts as well as for answers to their questions.

c. Read aloud to the students selections from a number of books suggested in the bibliography. After reading each selection, have children recall the new information they have learned. Record their responses in a web format.

d. Students might record their questions and answers on the bug shapes found on page 58 in this guide. Encourage students to record new and interesting facts they discover in their searches, too.

NOTES

Real-Life Laboratory

Collecting Insects

Collecting insects will give young children a sense of ownership of the *Bugs and Other Insects* theme unit. Select one or two books or articles on how to collect insects to share with the class before students go out collecting insects on their own. Demonstrate for the class how to handle an insect gently.

1. Reviewing Insect Characteristics

Review with students the characteristics of an insect, which are:

> six legs
>
> three body parts—head, thorax, and abdomen
>
> tough shell-like outer covering

2. Distinguishing Harmful from Harmless Insects

Acquaint children with harmless insects, such as ladybugs, caterpillars, grasshoppers, and crickets. Remind students that they should never pick up a spider or water insects, as they might be dangerous.

3. Collecting Insects

Have students work in pairs. Give each student pair a sheet of paper, a jar with a lid full of air holes, a small notebook, a pencil, and a magnifying glass, if possible. Take the children to a grassy area, a woodland, or a garden. Remind students to place some grass or leaves in their containers so that the insects they capture will have some food from their natural environment.

a. When students find an insect, they should do one of the following:

1. Pick up the insect very gently and place it in a container.

2. Slide the insect onto a piece of paper and then into a container.

3. Shake the insect off a low branch of a shrub or tree into a container.

4. Place a leaf or twig with an insect on it inside a container.

b. To trap bees and wasps, have students:

1. Make a funnel by rolling up a sheet of paper like an ice-cream cone and taping it together.

2. Cut off both ends of the funnel so that the smaller end fits into the top of the jar and the larger end extends out the top.

3. Put small pieces of fruit or meat into the jar and put the jar outside for a few days.

NOTES

4. Encouraging Observation

Encourage students to observe carefully what insects do and where they are found. Have students write their observations in small notebooks. A magnifying glass will help students see the small world of insects more readily.

NOTES

Classroom Observation Centers

Setting up and maintaining an insect garden or zoo will help children develop a sense of responsibility for bugs and other insects. These activities provide many opportunities for children to learn how to handle insects and will help them become careful observers as well. Begin the lesson by reading aloud a nonfiction selection, perhaps one on setting up an insect garden or zoo. Or, you might read about some of the specific insects collected by the class members.

1. Making an Insect Garden or Zoo

Encourage students to find a reading selection that provides directions for making an insect garden or zoo. If such a book isn't available or is inappropriate for the grade level, students can use the following steps:

a. Put all insects that have been collected into a terrarium or into individual jars.

b. Be sure to include the right kinds of food—fresh leaves, grass, twigs, and sticks from the area where the insects were found.

c. Put caterpillars in separate jars—one caterpillar per jar. Make sure the jar lid has holes in it so the caterpillar can breathe. Put leaves from the areas where the caterpillars were found in the jars as well. Caterpillars are very fussy about what they eat.

d. Each day, put fresh leaves, grass, and other food into the terrarium and jars.

e. Sprinkle a little water on the leaves each day to simulate dew, or put a wet sponge into the jar so the insects have water.

f. Keep the terrarium and the jars clean by removing all droppings and wilted leaves.

g. Keep the insects only for a short visit. Then return them to the grasses where they were found.

NOTES

2. Making an Ant Farm

NOTES

Make an ant farm using the directions from one of the reading selections collected for this unit. If an appropriate book is not available, students can do the following:

a. Sift dirt into a wide-mouth jar using a large sifter and a funnel.

b. Gather ants on a sheet of paper on which you have put crumbs or honey.

c. Put the ants into the jar.

d. Put a small wet sponge into the jar so the ants have something to drink.

e. Put bread and cookie crumbs and small seeds into the jar.

f. Watch ants outdoors to see what they eat or carry. Put the same types of crumbs that these ants eat or carry into the jar for the captive ants as well.

g. Put the jar in a dark place so the ants will feel like they are underground. Give the ants time to build some tunnels.

h. Take the jar into the light to observe the ants.

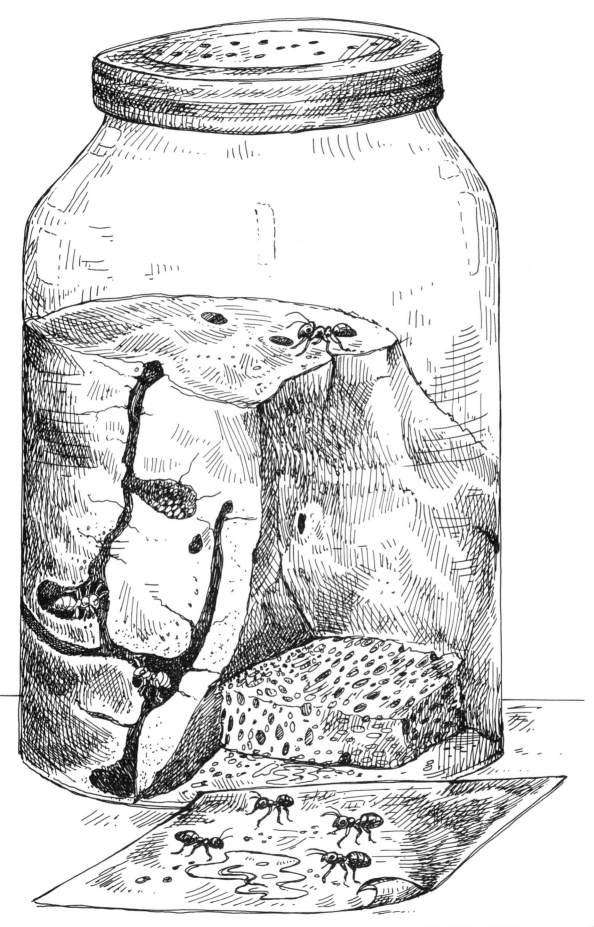

BUGS AND OTHER INSECTS 35

Observe and Record

By observing insects, children will make many interesting discoveries. They will also learn to ask questions about the natural world around them. Hand-held magnifying glasses will help the children observe more closely what the insects are doing. As always, begin each lesson by reading a nonfiction selection to the class to prepare students for the following activities.

1. Identifying Insects

Have students identify their insects and verify their identifications by using the nonfiction selections available in the classroom for this unit.

2. Focusing Observations

Provide time for students as a class, individually, or in small groups to observe the insects. To help students learn to focus their observations, pose "I wonder" questions. For example,

"I wonder what each of the insects is doing?"

"I wonder when and how an ant sleeps?"

"I wonder how _____ sees?"

"I wonder why the caterpillar eats all the time?"

"I wonder which insects don't harm people?"

"I wonder which insects look like their parents when they are babies?"

3. Recording Observations

Frequently record students' observations on a large chart.

WHAT WE SEE CHART

Name	How It Looks	What It Eats	What It Does	How It Moves

Cross-Curriculum Activities

Writing Arena

Writing helps children think about what they have observed and what they already know. It also helps them synthesize their thoughts and then communicate their ideas to others. Read a nonfiction selection to the class before beginning a writing activity. You might point out some sentences or paragraphs in the selection that are especially well-written.

1. Recording Observations

a. When children come back from collecting their insects (see page 30), take a few minutes to write down on a large sheet of paper their observations and reflections. Encourage children to describe the feelings they had while watching for insect specimens.

b. Have the children make small bug- or insect-shaped databooks and then record their observations of what is happening in the insect garden or zoo (see page 33). Remind students to date their entries.

2. Charting Metamorphosis

As a group, chart the metamorphosis of a caterpillar to a butterfly. Include the dates and the children's descriptions of what they have seen. Post this close to the jars of caterpillars.

3. Writing a Newsletter to Parents

Help children write a newsletter to parents.

a. Hold a class meeting to give children an opportunity to discuss what they are observing and learning about the different insects.

b. As a group, write a paragraph about their observations in the insect garden or zoo (see page 33).

c. Have children individually or with a partner draft a paragraph describing something they have learned about one of the insects. Hold a writers' conference with two or three children in each group. Invite each author to read his or her paragraph aloud. Help children with positive suggestions for revisions. Suggestions for revising and editing are found on butterfly-shaped writing helpers on pages 59-61.

d. Compile all paragraphs into a newsletter for parents, the principal, and students in other classrooms. Invite readers to respond to the articles in the newsletter.

NOTES

Art All Around

Art activities help children visualize the concepts they have been reading about and discussing. Frequently, through art, children will want to review something they have read, heard, or observed so that their artwork is as accurate as they can make it. You'll want to have several resources available for children to refer to. Select a book or magazine article with good photographs or illustrations to read aloud to the class before beginning any activity or activities.

42 BUGS AND OTHER INSECTS

1. Habitat Mural

Create a mural of the habitat of the insects the students have been studying. For texture, use a variety of materials, such as:

- construction paper
- bugs made from clothespins and tissue
- crayons
- markers
- paints
- tissue paper

NOTES

2. Moveable Insects

Construct moveable ants, crickets, or grasshoppers using the patterns provided on pages 62-64. Have the children follow these steps:

a. Select an insect pattern.

b. Color the insect parts to match the real insect.

c. Cut out the insect pieces and fasten them together with brads.

d. Display in the classroom.

3. Butterfly Mobiles

To make butterfly mobiles, provide clothespins and patterns for wings. Have the children follow these steps:

a. Decorate a clothespin with crayons, markers, or paint.

b. Color two sets of wings.

c. Tape the wings to both sides of the clothespin.

d. Tie a thread to each butterfly so it can hang from the ceiling.

4. Magnetic Ants

NOTES

The children will enjoy making and playing with magnetic ants. Provide brads and black, brown, or red construction paper. Help the students as needed with the following steps:

a. Draw and cut out three oblong pieces about the size of a nut for body parts.

b. Glue the body parts together.

c. Punch three holes (for brads) into the middle body part of each ant.

d. Put a brad through each hole and open the brads for the six legs.

e. Have children punch another hole in one end of the ant.

f. Place a brad so that the head of the brad is on the underside of the ant. Bend the ends of the brad to make feelers.

g. Place several ants on a paper plate. Use a magnet on the bottom of the plate to make the ants move.

BUGS AND OTHER INSECTS

The Drama Scene

Drama is a positive, fun, and fulfilling way of learning in which children can practice predicting, planning, organizing, and problem solving. Before beginning one of the following drama activities, read aloud a nonfiction selection relating to the activity or activities.

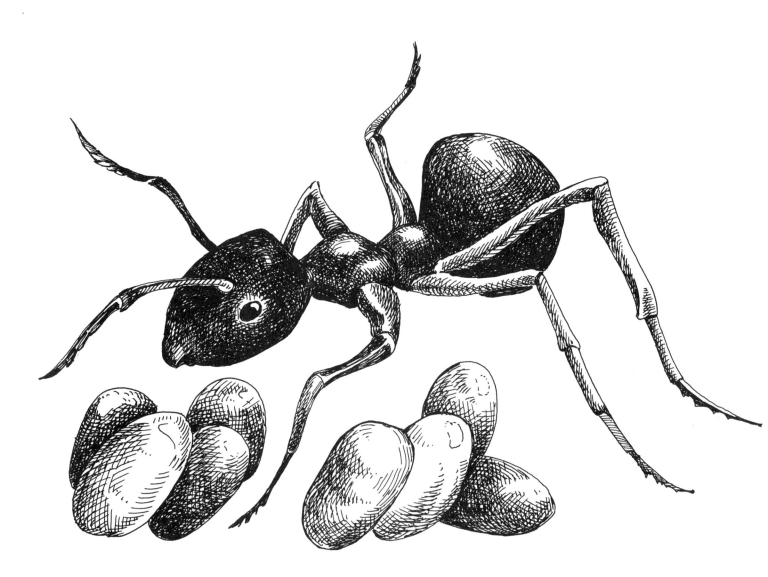

1. Imaginative Play and Movement

NOTES

Give children their own space for movement by asking them to stretch out their arms, then turn and stretch out their arms again. Ask children to pantomime all their movements within the space around them. Create action by having students follow the pantomime directions as you give them.

Narrative Pantomime—
From Caterpillar to Butterfly

"You are a caterpillar. You are on a twig. You are hungry. You just keep eating leaves and more leaves. Carefully take a bite out of a leaf. Use your strong jaws to cut a hole in the leaf you are eating.

"You are growing, growing. Your skin feels very tight. It's beginning to split. Slowly you are crawling out of your skin. It's hard to crawl out of your skin, but you are doing it.

"You keep on eating. You are getting larger and larger. You have gotten so big that you must shed your skin again. Very slowly crawl out of your skin again.

"Now you are getting very still. You are forming a pupa or chrysalis. Look for a firm tree branch. Spin a pad of silky threads underneath a leaf. Now spin a small knob on the silky thread. Hook your back feet onto the knob. Your skin is splitting for the last time. Your new skin is soft at first. Then it gets harder and harder. Now it is very hard. You are very quiet as you hang upside down.

"A long time has passed and your chrysalis is beginning to open. Push your wings out very carefully. Slowly move your new wings up and down so they can dry. You are now a beautiful butterfly. Fly a short way to make your wings strong. Fly again to a flower for sweet nectar."

> *Narrative Pantomime—*
> *Worker Ants*
>
> "You are a worker ant. There are many rooms in your house. Go into the large room or chamber to see the eggs in their cocoons. Now go to the food storage rooms.
>
> "It's time for you to go look for food. Look for crumbs, jam, ripe fruit, or sugary foods. Take some food back to your home. Feed the young ants."

2. Describing an Insect's Life Using a Storyboard

a. Encourage each student to make a storyboard from a 9" x 12" piece of cardboard or ask students to create a scene showing an insect habitat on the front of the board. Provide construction paper, crayons, and markers.

b. Have students construct an insect puppet from cardboard or oaktag. Students can make a stick puppet by gluing a tongue depressor to the back of the puppet shape. Another idea is to glue a magnet to the puppet back and move the insect around with another magnet behind the storyboard.

c. On the back of the storyboard, have children glue a pocket for storing the insect puppet.

d. Have the children describe to others the life of the insect they have been portraying. Students might include where the insect lives, what and how it eats, how it grows, and how it stays away from its enemies.

Front

Pocket

Back

BUGS AND OTHER INSECTS 49

Fingerplays and Poetry

*F*ingerplays and poetry give children opportunities to play with the rhythms of the language. The fingerplays and poetry suggested in this guide, for the most part, are based on specific insects. Before beginning the lesson, read aloud nonfiction selections related to the insects in the activities.

1. Fingerplays

"Here Is the Beehive" and "The Caterpillar" are two fun fingerplays you'll find in Marc Brown's book *Hand Rhymes* published by E. P. Dutton. Provide several opportunities for the children to do the fingerplays as a group or as individuals.

2. Chanting

Read and chant the poem "We Like Bugs" by Margaret Wise Brown. Here are some suggestions for chanting.

a. Divide the class into two groups. Have the first group read the first line. Then have the second group read the second line. Continue with the groups reading every other line.

b. Start reading the poem very quietly, gradually increasing the volume. Then do the reverse and see which version the class prefers.

c. Assign each line to a different person or pair of students.

3. Choral Reading

Another poem that is good for choral reading is "Grasshoppers" from *Joyful Noise* by Paul Fleischman, published by Harper & Row. When the children think they are ready, you might have them read "Grasshoppers" aloud to another class as well.

NOTES

Cooking Makes It Memorable

*C*ooking activities add the sense of taste and smell to the learning experience and help children form lasting impressions of what they have learned. Begin the lesson by reading a book or magazine article on butterflies aloud for the first activity, or read a general selection on simple insects.

52 BUGS AND OTHER INSECTS

1. Making Celery and Pretzel Butterflies

Help children make butterflies from stuffed celery sticks and wing-shaped pretzels. Cut celery sticks about as long as the pretzels and stuff with peanut butter or soft cheese. Carefully insert a wing-shaped pretzel on each side. If necessary, top with a little more filling to hold the pretzels in place. If children wish, they might ice the wings and sprinkle with cake decorations.

2. Ants on a Log

Help children make ants on a log using stuffed celery and raisins. Cut celery sticks about five inches long. Fill with peanut butter. Place raisins on the peanut butter filling.

3. Baking Bug-Shaped Cookies

a. Give each child a slice from an unbaked cookie dough roll.

b. Have children shape the dough with their fingers to form different kinds of bugs. Remind children that lumps of dough will bake together.

c. Decorate the bugs with:
   - nuts
   - chocolate or butterscotch chips
   - dry cereal
   - raisins
   - cinnamon candies

d. Bake at 375 degrees until cookies are golden brown.

NOTES

Social Studies from Here to There

To help children become acquainted with the globe and to further their interest in the world beyond their own environment, help them trace the migration route of the monarch butterfly. Begin by reading aloud a book or magazine article about the monarch butterfly migration. Try to have several resources on monarch butterflies available for the children.

Butterfly Migration

Discuss what migration is with the children. Then select any or all of the suggestions provided below, depending on the needs and interests of your students.

a. If you live east of the Rocky Mountains, trace a route from your area to the mountains just north of Mexico City.

b. If you live west of the Rocky Mountains, trace a route down the coast of southern Oregon, to southern California, or to the Baja Peninsula of Mexico.

c. Introduce children to the mileage page in an atlas and help interested students add up the distance the monarch butterflies might fly for the winter.

d. Have students speculate why the monarch butterfly migrates to Mexico. Discuss the location of the equator and the warmth of the sun at that location.

NOTES

Science Sector

*T*he following activities help children understand how different members of the animal kingdom, including insects, depend on one another. A nonfiction book or magazine article on food chains would be excellent to read aloud before doing either of the activities suggested in this section.

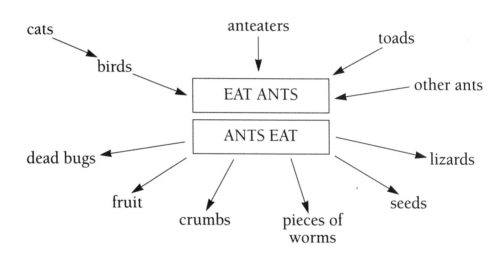

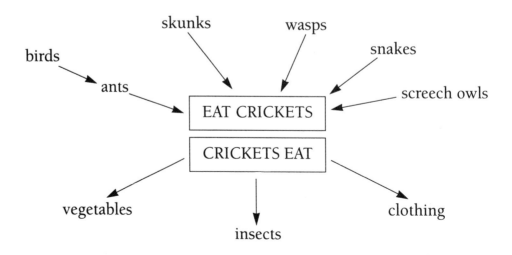

1. Developing Food Chains

As children are observing and reading about insects, sponsor a group meeting. Discuss with children what insects eat. Develop several food chains, such as the ones shown at the left.

2. Studying Food Chains

Use a ball of yarn to demonstrate how a food chain works.

a. Hold on to the end piece of a ball of yarn as you hand the ball of yarn to a student. Have him or her name a food that an insect eats, such as crumbs.

b. The student hands the ball to a second student who then calls out the name of an insect that would eat the food identified by the previous student.

c. That student, in turn, passes the yarn to a third student who names an animal that would eat the insect identified by the second student.

d. As the food chain develops, the ball of yarn unwinds, and students begin to visualize the food-chain relationships in nature.

NOTES

Membership Cards for a Bug Club

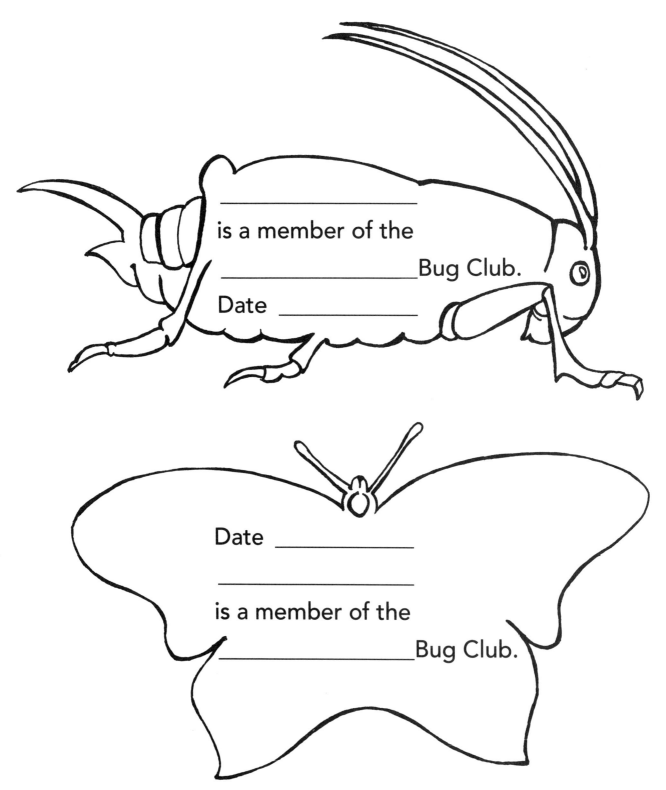

_____
is a member of the
_____ Bug Club.
Date _____

Date _____
_____
is a member of the
_____ Bug Club.

Writing Helpers

Directions: Make a copy of the writing helpers for each student. Laminate for durability. Assemble each set in order and fasten together with a ring. Encourage children to use their writing helpers whenever they are working on a writing activity.

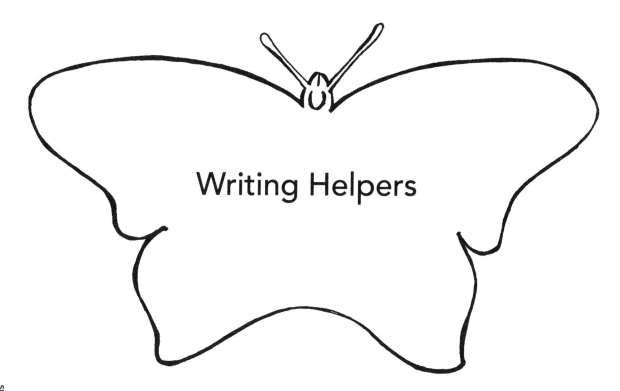

Share

1. Read your draft to a partner.
2. Ask your partner if your draft makes sense and if it sounds right.

Listen and Revise

1. Listen to your partner's ideas.
2. Let your partner help you.

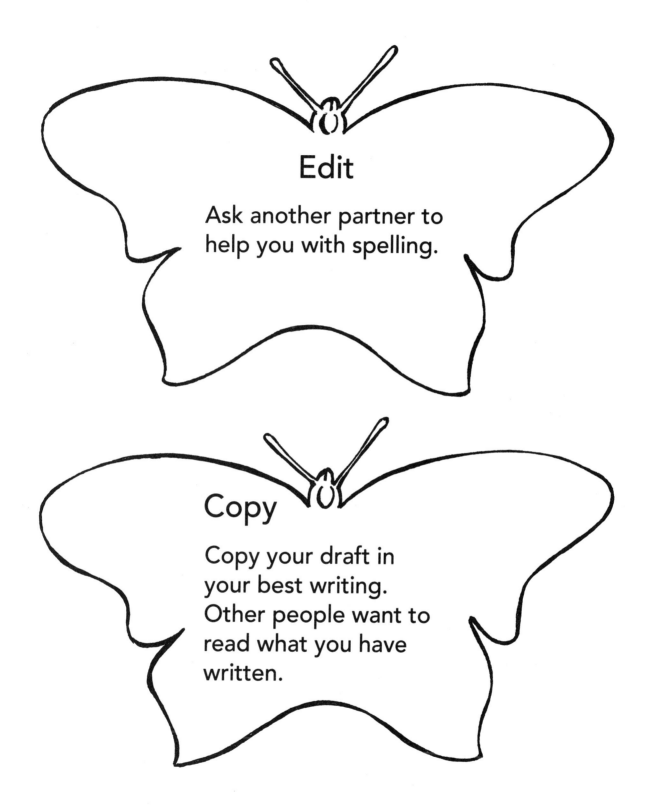

Edit

Ask another partner to help you with spelling.

Copy

Copy your draft in your best writing. Other people want to read what you have written.

Ant Pattern

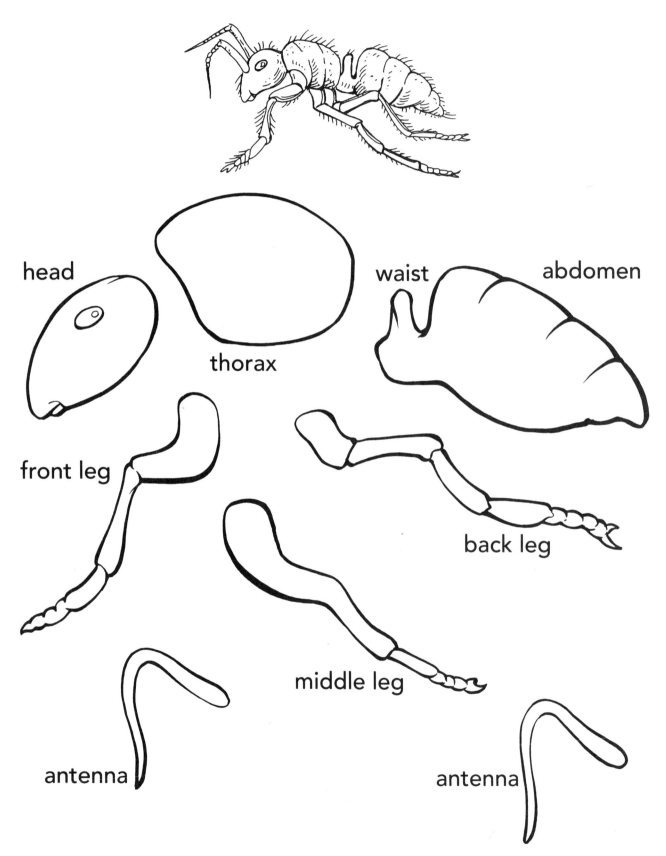

Cricket Pattern

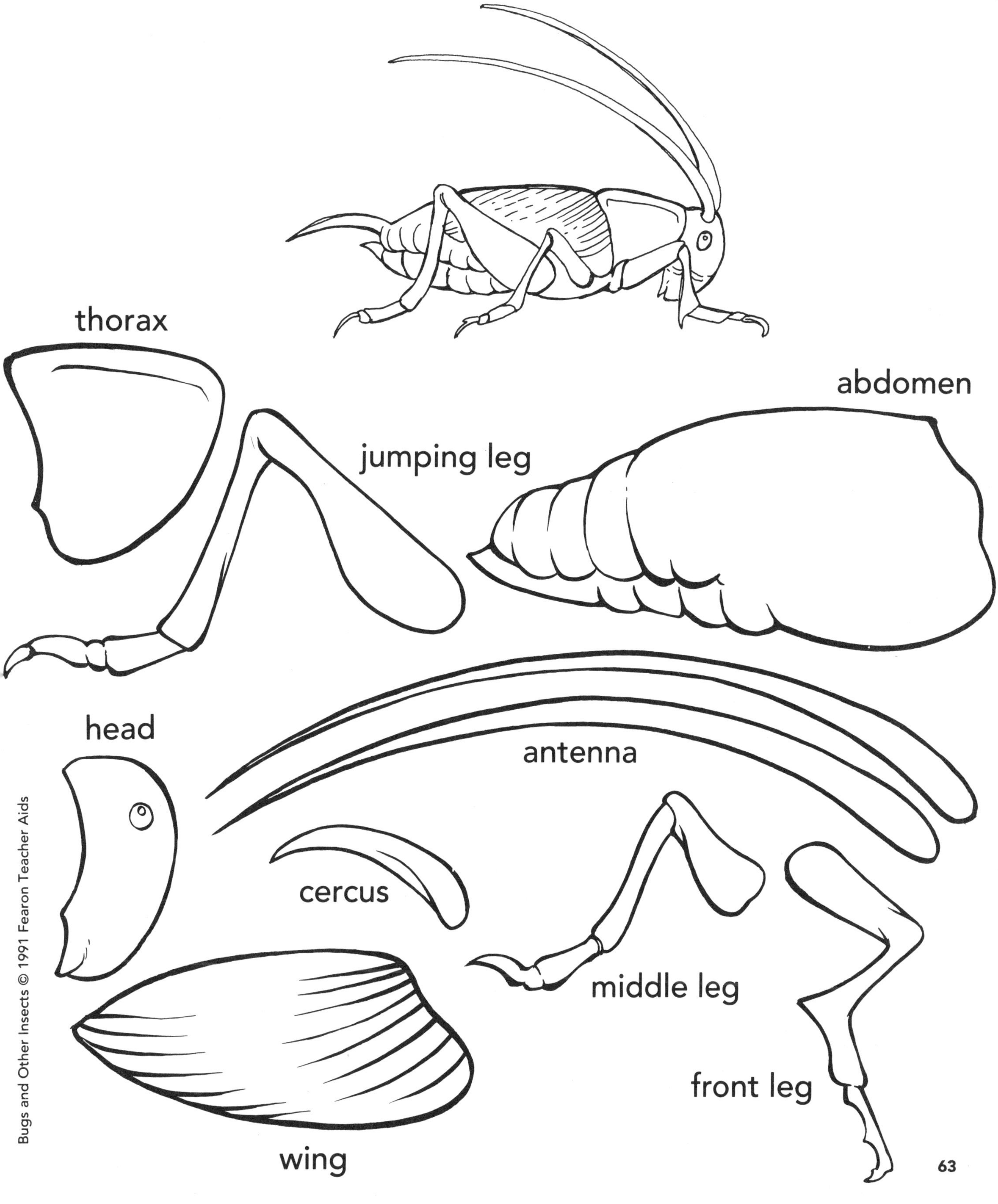

thorax

abdomen

jumping leg

head

antenna

cercus

middle leg

front leg

wing

Grasshopper Pattern

thorax

jumping leg

antenna

head

abdomen

front leg

wing

front leg

wing

middle leg